THE FUTURIST VET

SCIENCE FICTION VETERINARY HEALTHCARE IS ALREADY HERE

DR. GORDON ROBERTS BVSC MRCVS

TABLE OF CONTENTS

FOREWORD

Dear Reader,

The fourth industrial revolution has begun and medicine as we know it is changing beyond recognition.

From 3D-printed organs and autonomous robotic surgeons to genetic engineering and multiple in-body sensors, the things we previously thought belonged in the pages of science fiction novels have become science fact.

It is such an exciting time for the veterinary profession as these 'bleeding-edge' technologies and mind-blowing medical discoveries begin to change the lives of companion animals too.

So, perhaps you'd better sit down as I reveal to you the future of veterinary medicine...and if this taste of the future leaves you wanting more, sign up to my regular newsletter at futuristvet.com.

Dr. Gordon Roberts

The Futurist Vet
December 2016

INTRODUCTION

"It is not the strongest of the species that survives, nor the most intelligent, but the one most responsive to change.
– Charles Darwin

In the London Science Museum there is on display a doctor's bag dated 1890-1930 that belonged to John Hill Abram, a Liverpool physician who later became Professor of Medicine. What is most striking about this century old bag of physician's equipment is how similar it is to the medical bags still in use today. Nothing much has changed in one hundred years of healthcare.

The doctors' bags of the early 1900s contained syringes, thermometers (albeit mercury ones), tongue depressors, stethoscopes, ophthalmoscopes for examining the eyes and sphygmomanometers for measuring blood pressure. Today the thermometers and blood pressure monitors are electronic and digital but the rest is largely the same. This is a reflection of the medical profession in general.

However, we are at the dawn of a technological revolution, which will see a complete transformation of both human and veterinary medicine. Medical innovator Lucien Engelen says that after the invention of the steam engine, introduction of mass production and the emergence of the internet we have reached the fourth industrial revolution – the Internet of Things – and this will change the face of healthcare forever.

In every aspect of our lives we are becoming increasingly more connected and the medical profession has to change, and change quickly, to keep up with these emerging new technologies.

The doctor's bag that we are all familiar with is soon to be replaced with a digital version, comprised of a smartphone or tablet computer, wearable sensors, other connected items and condition-specific apps.

There will be apps to measure every aspect of a patient, from blood pressure and heart rate, to blood sugar level and urine analysis. There will also be smartphone plug-ins for ear and eye examinations, as well as imaging. And that's just for starters!

Up until recently dogs and cats didn't live much longer than their ancestors. Society has changed, placing greater importance on companion animals as part of the family and therefore medical care for pets has become more and more important.

The rise in pet insurance reflects this and the belief is that, if a pet is ill or injured, the cost of treatment should not be a barrier to them receiving the best possible care. The reasons for pets being euthanised just a few years ago have become treatable with modern veterinary medicine and, with insurance to pay for it, the best vet care is available to all.

Where human medicine has gone, veterinary medicine has followed and today our pets have access to cutting-edge technologies including MRI scans, chemotherapy, key-hole surgery, prosthetics and DNA testing. With the imminent revolution in human healthcare we can look forward to an explosion in 'bleeding-edge' technologies such as 3D printing, gene therapy, and robotic surgery being offered to our pets in the near future.

But this is not all. The way in which we approach healthcare will be turned on its head from what is currently a reactive, sickness-led approach to following the four Ps as outlined by American biologist Leroy Hood – Predictive, Preventative, Personalised and Participatory.

The healthcare and veterinary care of the future will place more focus on the prevention of disease, tailoring treatments to individual patients with an understanding that one size does not fit all and engaging patients (or in the case of pets, their owners) to take control of their health.

In 1900, 80% of veterinary work was with horses but within a decade that dropped to just 10% due to the launch of the Ford Model T motorcar. The veterinary profession at that time had to adapt to disruptive changes in society and this is happening again now with the fourth industrial revolution.

Read on to find out how completely new ways of thinking, jaw-dropping technological advancements and even the stuff of sci-fi novels are going to transform veterinary medicine beyond belief.

NEW MEDICINE

"We can change our mindset from 'sickcare' to 'healthcare' by shifting from a reactive era of medicine to an era of medicine that is proactive, preventative and continuous care."
- Dr. Daniel Kraft

The way in which we approach human medicine is changing dramatically and this will influence the future of medicine for our companion animals too. The body is no longer seen as a collection of separate organs and processes but as a finely tuned system requiring a holistic view to maintain it.

Also, the similarities between species mean that human and veterinary medicines are often very similar and any advances in one field should be used to benefit the other. Animals have long been used for medical research and it's about time that our pets reaped the benefits of this through improved medical care for themselves.

One Health

The term "One Health" is fairly new but the concept of a universal way of viewing healthcare dates back to ancient Greece when the environment was first recognised as being a contributing factor to human health. With the discovery that diseases could be passed from animals to humans - such as malaria being spread by mosquitos - human, animal and environmental health became inextricably linked.

German physician Rudolf Virchow was the first to use the term 'zoonosis' in the late 19th century to describe infectious diseases in animals that can naturally be transmitted to humans. He is also quoted as saying, "...between animal and human medicine there are no dividing lines – nor should there be."

The One Health Initiative is a movement that was set up to promote the equal collaboration of physicians, veterinarians, dentists, nurses and other scientific and environmentally disciplines for the benefit of all. Vets are uniquely trained in comparative biology and see many different species of animal in their day-to-day work. They play a vital role in preventing new diseases in humans, as well as potential plagues, and it's safer, cheaper, healthier and more effective to identify a disease in an animal before it appears in humans. It's a vet's ability to link animal science to human wellbeing that is advancing food production and safety, and providing a critical defence from global pandemics.

Penn Vet, part of the University of Pennsylvania, is the only veterinary school developed in association with a human medical school. One of its focuses is on what it calls translational medicine, working with diseases that affect humans and animals with the aim of benefitting both.

For example, osteosarcoma bone cancer tumours found in dogs are very similar to those found in children. The research being done by Penn Vet on canine osteosarcoma will ultimately help children with bone cancer too.

A Penn Vet team has been trialling a bacterial vaccine aimed at kick-starting the body's immune system response to tumoirs in dogs. The vaccine has been found to stimulate the immune system safely and the effects have prolonged the lives of the patients. Based on these results the vaccine is now being taken into human clinics where it will be evaluated in people with certain types of cancer.

The Penn Vet Shelter Canine Mammary Tumour Programme provides care to shelter dogs with mammary tumours, treating them and therefore making them more adoptable.

The programme includes screening tumour staging to determine how advanced a tumour is, tumour surgery and regular follow-up exams. Spontaneous tumours in companion animals represent a previously untapped resource in cancer research in general and offer a new way of studying cancer in a natural setting.

The research may have direct applications to cancer in humans as dogs with mammary tumours and women with breast cancer are very similar in terms of biology, dietary risk factors, clinical behaviour and hormones. Treatment failures resulting in recurrence and eventual death remain the major obstacles in breast cancer treatment for both dogs and women.

Because dogs typically have 10 mammary glands and often develop tumours in several glands at the same time they present a unique research opportunity, enabling scientists to study tumors that are at different stages of development in the same animal.

Professor Noel Fitzpatrick, also known as the Bionic Vet due to his cutting-edge work with pet prosthetics, is the founder of The Humanimal Trust, which works on building closer working relationships between vets and doctors.

The Humanimal Trust believes it is wrong that an animal may not go on to benefit from a procedure, device, therapy or drug they were used to develop. It believes we should move towards the one medicine view that human and veterinary medicine are dependent on an overlapping collection of biological characteristics, technologies and research discoveries.

Professor Fitzpatrick is quoted as saying, "I passionately believe that a framework among human and veterinary clinical colleagues that is open, honest and most of all organised, will bring us closer to a just society where all patients win – not one species at the expense of the other, but rather saving a life to help save another."

Functional Medicine

Functional Medicine (FM) is a holistic approach to healthcare that focuses on interactions between the environment and gastrointestinal, endocrine and immune systems. It is often described as the clinical application of systems biology, which thinks of the body in the same way as you might a computer.

Primarily FM is a medicine of cause, not symptoms, and is concerned with connecting the dots with a more personalised medicine.

More than 75% of healthcare costs in the US are due to chronic conditions, such as heart disease, diabetes and arthritis, and these costs are rising at an unsustainable rate.

Supporters of FM, including Dr. Mark Hyman from the Cleveland Clinic who is a leader in this field, believes the current healthcare system fails to address the causes of chronic disease by not taking a holistic approach.

Dr. Hyman says the treatment of chronic disease is only possible when the cause of the disease is identified and treated, rather than simply managing the symptoms.

In the veterinary world chronic disease has risen drastically over the past few years too. Conditions such as obesity, arthritis, kidney disease, thyroid disease and heart disease are becoming more common.

A report compiled by Banfield Pet Hospital in 2012 captured and analysed medical data from two million dogs and nearly 430,000 cats.

It showed that arthritis alone had increased by 38% in dogs and 67% in cats.

FM works on the basis that chronic disease is influenced by many different factors including lifestyle, genetics, the environment and diet so these are the areas that need to be looked at when treating our pets.

Dr. Hyman is quoted as saying, "Real food is the most powerful tool we have to fight illness, it's a better drug than drugs."

Precision Medicine

There is an emerging field for disease treatment and prevention in humans that is making medicine more personalised. This Precision Medicine (PM) takes into account an individual's genes, environment and lifestyle, and could be used just as easily with pets now that we have commercial genetic sequencing available. It is no longer a case that 'one size fits all' when it comes to treating disease.

In 2015 President Obama launched the Precision Medicine Initiative – a bold research project that is looking at improving health and treating disease in a new way. President Obama is quoted as saying, "Doctors have always recognised that every patient is unique, and doctors have always tried to tailor their treatments as best they can to individuals. You can match a blood transfusion to a blood type – that was an important discovery. What if matching a cancer cure to our genetic code was just as easy, just as standard? What if figuring out the right dose of medicine was as simple as taking our temperature?"

Advances in PM have already led to new discoveries and treatments tailored to an individual's genetic make-up, which is helping transform the way we treat diseases such as cancer. Researchers at the National Cancer Institute in the US hope to use PM to find new, more effective treatments for various kinds of cancer, looking at genetics and the biology of the disease. One aspect of PM is pharmacogenomics, which is the study of how genes affect a person's response to particular drugs. In the future drugs for people and their pets will be tailored to variations in their individual genes.

Integrative Medicine

There has been an increase in the popularity of integrative medicine in recent years as a response to social changes. Healthcare for humans has moved away from sole practice primary care to large healthcare systems that some patients feel lost in. Integrative medicine is all about care that takes account of the patient as a whole and their lifestyle, while also emphasising the relationship between patient and practitioner.

The most obvious difference with integrative medicine is the longer consultation times and the use of therapies such as acupuncture and homeopathy.

Many of these alternative therapies date back thousands of years so they have not been subjected to the clinical trials required by today's medical standards.

Due to general skepticism surrounding these therapies trials are now being funded with the results published in conventional medical publications.

This is a move away from the 'conventional versus alternative' view to an integrative approach that uses both medicines in partnership with each other. Integrative medicine only uses complementary therapies for which there is some high-quality scientific evidence of safety and effectiveness.

There is clear evidence that integrative medicine is becoming part of current mainstream medicine and this is also expected to become the case in veterinary medicine over the coming years. Treatments that may be offered alongside conventional treatments as part of integrative medicine include:

• **Traditional Chinese Veterinary Medicine** – this is comprised of four branches, which are acupuncture, herbal medicine, food therapy and tui-na (a form of medicinal massage).

- **Acupuncture** – the stimulation of certain points of the body using thin, solid, metal needles. *(See image)*

- **Homeopathy** – a system of remedies that work on the principle that 'like cures like'. Homeopathic remedies are derived from substances that come from plants, minerals or animals and are formulated into sugar pellets that can be placed under the tongue.

- **Phytotherapy** – herbal medicine that is based on scientific or medical evidence so its products are pharmacologically active, similar to conventional pharmaceutical drugs.

- **Light therapy** – harnessing the healing power of red and infrared light at specific wavelengths and frequencies to reduce swelling and inflammation, stimulate the immune system and relieve pain.

THE INTERNET OF ANIMALS

"The Internet of Things has the potential to change the world, just as the Internet did. Maybe even more so."
- Kevin Ashton, who coined the term 'Internet of Things'

The number of smartphone users in the world is forecast to reach 2.5 billion in 2019 (around a third of the population) but phones are just one of a growing surge of devices that are now connected to the internet.

Even the most mundane of household appliance is going to be 'smart' in the near future with experts saying that by 2017 nine out of ten will be Wi-Fi controlled. From internet-enabled washing machines that you can operate remotely to smart refrigerators that can tell you when food is going off, it won't be long until all our appliances, and even our pets, are connected.

This network of connected devices is known as the Internet of Things and by 2030 it is predicted there will be 500 billion devices and objects connected to it. A growing number of these connected devices are actually being used by pets, which is why the Internet of Animals has become big business.

Innovators are beginning to realise the almost endless potential of connected technology for companion animals. A whole host of start-ups have joined the Internet of Animals, each with a slightly different take on connecting people more closely with their pets and improving pet health. Raymond McCauley, a scientist, engineer and entrepreneur working at the forefront of biotechnology believes that toilets will become the most networked appliance in the home of the future.

He says that one day toilets will be used to monitor our health and diets, adding that they could even provide tips such as telling users to "stop eating chorizo".

Pets seem to be one step ahead when it comes to smart toilets as there is already a device that turns any cat litter tray into a

verizon 📶   9:42 AM   25% ▮

tailio

Cats

Maui
12.1 lbs
Last visit 5:16 AM on Aug 31
⚠ Maui has 15 alerts

Mia
12.7 lbs
Last visit 7:47 AM on Aug 31
⚠ Mia has 11 alerts

Mollie
9.4 lbs
Last visit 10:27 PM on Aug 30
⚠ Mollie has 44 alerts

Devices

📊   〰   ✉   ○ ○ ○
board   Activity   Inbox   More

smart internet-enabled monitor. The Tailio device sits under a cat's usual litter tray and works in the background to monitor their health. It can even monitor multiple cats by learning the characteristics of different users.Tailio collects data on a cat's weight, amount of waste, frequency of visits and behaviour in the litter box.

Powerful analytics then learn a cat's behaviour from the data and Tailio is able to watch for trends and changes such as weight loss and subtle changes in elimination behaviour. The cat owner installs an app on their smartphone which enables Tailio to send alerts if something needs their attention, such as their cat losing weight or using the tray more than usual.

Smart pet feeders that can control portion size and monitor frequency of use were initially seen as a bit of a gimmick but as they have become more sophisticated it's clear they have a role to play in improving pet health.

Feeders such as the SmartFeeder from Petnet are now more personalised, delivering food based on a pet's age, weight and level of activity to reduce obesity. A mobile app allows an owner to monitor their pet's status, control the feeding schedule and assign correct portions. Pets can be fed remotely and food can be re-ordered through the app. The SmartFeeder is able to learn about a pet over time and can calculate the calories they are consuming.

The Obedog ProBowl is another connected feeding device that aims to make life easier for owners while improving pet health. The ProBowl weighs and measures food as it is filled and changes colour when the right amount has been dispensed, based on a pet's individual requirements.

An owner can see clearly via their smartphone app when their pet last ate and how much food was taken. The really clever thing about the ProBowl is that you will never run out of food again as it's connected to Amazon's Dash Replenishment Service. This means the ProBowl automatically tracks the amount of food used and re-orders before your stocks run out. So, that's one less thing to add to your weekly shopping list! The company is now working on a version of the ProBowl especially for cats.

Separation anxiety can be a common problem for dogs that are left at home alone during the day, but would they feel more reassured if their owners were able to speak to them and even give them treats when they're not there?

The PetBot uses what it calls petificial intelligence (artificial intelligence focused on pet interaction and care) to bring owners new pet initiated experiences. The PetBot is a Wi-Fi treat dispenser and webcam in one that aims to keep pet and owner connected at all times.

Using the PetBot smartphone app an owner can watch their pet via a live stream, speak to them through their phone, dispense treats remotely and even get sent pet 'selfies' during the day when the PetBot's facial recognition software spots their pet is in front of the camera. It even uses clever algorithms to

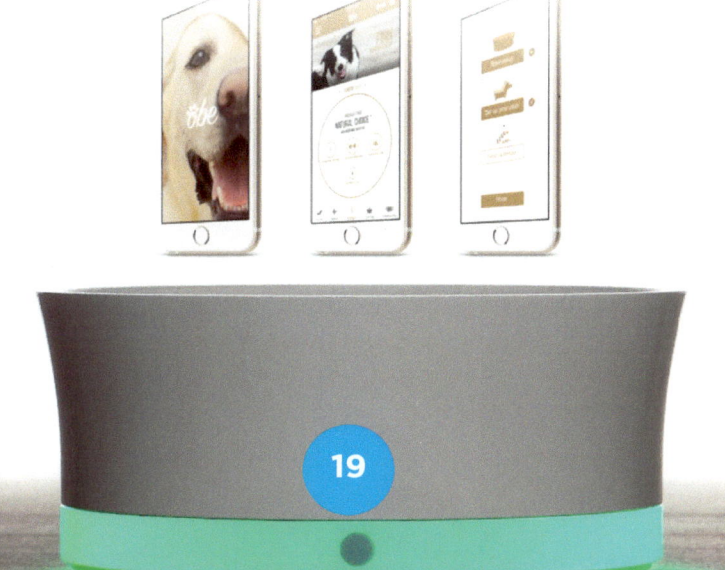

recognise barking and can send an alert to an owner if their pet sounds stressed.It is not just the appliances a pet is using in the home that have become connected but the pets themselves, due to the rise in wearable technology. These 'wearables' include GPS trackers, activity monitors and small webcams that give owners a pets' eye view of life.

A report by independent market research company IDTechEx forecasts that the global market for animal wearable technology will reach $2.6 billion in 2025. This includes livestock and wild animals that are becoming increasingly connected too. The areas in which IDTechEx sees this technology benefitting animals are: identification and tracking, safety and security, behaviour monitoring, behaviour control, medical diagnosis and even medical treatment.

The gaining in popularity of human fitness trackers, such as the Fitbit, has prompted similar wearable devices for pets. Voyce is a health monitor and wellness management system that has been developed with biomedical engineers and Cornell University College of Veterinary Medicine.

It uses non-invasive sensors built into a collar to gather information about a pet, including heart rate, respiratory rate, activity, calories burned and quality of rest. With this information Voyce can create a vital sign baseline so that owners are easily able to detect any luctuations in their pet's health.

Whistle is a wearable activity monitor that can also track a pet's location on demand using GPS technology. An owner can receive notifications via the mobile phone app or text alerts if their pet leaves a designated 'home' area and they are able to track them anywhere in the USA using a map view.

It is not just companion animals who are benefitting from wearables as there are racehorses with under-saddle sensors monitoring their speed, stride length and heart rate, which have proved to not only be a valuable training tool but also a way to prevent overtraining. If a horse's heart rate is unusually high it is likely to be a sign that he is unwell, even if he isn't displaying any discernible symptoms, and training at this time could be dangerous to his health.

Working dogs, service dogs and assistance dogs have very important information that they need to pass on to their handlers or others in an emergency but they are very limited in the ways they can currently do this. The Georgia Institute of Technology identified this as an issue and set about helping these dogs with occupations find a voice.

A team at the Institute has developed a communication vest for dogs with sensors that can be tugged or bitten by the wearer to activate certain alerts. This can be in the form of sending a text message, alerting the emergency services with GPS coordinates or setting off an audio recording that says "Excuse me, my owner needs your attention."

The future of pet wearables actually looks like it could be fairly short-lived with a move being made towards pet inside-ables. Ingestible technology can already be found in agriculture and there are currently dairy cows with tiny sensors sitting in their stomachs monitoring their health.

Whenever they pass a Wi-Fi point their data is uploaded to a database and this constant monitoring enables farmers to identify potential health problems quickly, reducing any unnecessary suffering for the animal and expense for the farmer.

As the cost of ingestible technology decreases there is a definite market for such tech in the pet sector. As our pets can't tell us when they're unwell - and some are particularly adept at hiding any signs of illness - who wouldn't want an alert sent to their smartphone when their dog or cat is feeling poorly?

Like a 'check engine' light on a car, early intervention at this stage could prevent future suffering or life-threatening situations. Inside-ables will make constant health monitoring possible in the future and we'll look at this in more detail in chapter seven.

UNZIPPING GENES

"Our genes are our predisposition, but they're not necessarily our fate."
– David Ornish MD, founder of Preventative Medicine Research Institute

A genome contains all the information needed to build and maintain an organism. It is made up of genes that contain a particular set of instructions in the form of DNA. In humans, a copy of the body's entire genome is contained in all its cells except those with no nucleus, such as red blood cells.

The Human Genome Project was set up in 1990 to provide a complete and accurate sequence of the DNA that makes up the human genome and to identify the estimated 20,000 to 25,000 human genes. This project was a success but it took over a decade to complete at a cost of nearly $3 billion.

The cost of sequencing a human genome has fallen drastically in the last decade from $100 million in 2001 to just $200 today. 23andMe was the first company to offer commercial genome sequencing at an affordable price, which has made it available to the masses.

Anyone can order a kit from 23andMe online, provide a saliva sample by post and receive an in-depth report telling them if they are a carrier of certain inherited conditions and how their genes play a role in their well-being and lifestyle choices.

Raymond McCauley, a scientist, engineer and entrepreneur working at the forefront of biotechnology predicts that in 2018 the cost of sequencing a human genome will be less than $20 and by 2022 it will be as cheap as flushing a toilet.

In 2005 researchers at the Broad Institute of MIT and Harvard published the canine genome sequence from a female boxer dog. Despite the dramatic differences in the physical appearances of dog breeds they share large segments of their DNA so, any sequencing for a particular breed is likely to apply to others.

Knowledge of the canine genome in combination with the human genome is helping narrow the search for genetic contributors for diseases such as cancer. Many of the cancers that affect dogs are biologically very similar to those affecting humans and any advances made in the treatment of these cancers will benefit both species.

More recently work has been done to sequence the feline genome too with the hope of being able to eradicate genetic diseases that affect cat breeds. Domestic cats possess more than 250 naturally occurring hereditary disorders, many of which are similar to those affecting humans. The domestic cat also makes an excellent model for human infectious diseases, particularly HIV/AIDS as the feline immunodeficiency virus (FIV) is a genetic relative of HIV, which causes AIDS.

Just as 23andMe has opened up genome sequencing to anyone there are now companies offering the same for pets. Embark is an advanced DNA test that can reveal the actual breeds in a dog's ancestry as well as screening for more than 160 inherited diseases. A dog owner can simply buy an Embark kit online, post a cheek swab from their pet to the Embark laboratory and unlock the secrets in their dog's genes. By tracking over 200,000 genetic markers Embark can provide an overview of both genetic disease and inheritable traits.

Embark looks for genetic variants associated with particular diseases and will inform an owner if their pet has a higher than average chance of developing one of them. Embark can also let breeders know if a dog is likely to pass any diseases-associated gene mutations on to their pups.

It is hoped that the information provided to owners will help them to better understand their pets' health, plan for the future and provide the best possible personalised care.

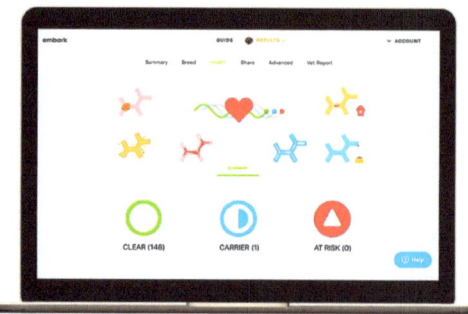

In 2016 Embark, along with Cornell University College of Veterinary Medicine, published the largest ever study of dog genetics using samples from more than 10,000 dogs worldwide.

Researchers successfully identified segments of the genome associated with shedding, body size and several inherited diseases. Following this they were able to predict the size of a dog with 90% accuracy.

Genetic engineering

Now that we understand the genomes that make up humans and other animals we can not only see the blueprint for these organisms but also find ways to alter them. Genetics has transformed biology into something that can be edited, just like a computer programme, with endless possibilities. This does raise the question of ethics and just how far should we go.

CRISPR-Cas9 is a genome-editing tool that has the potential to revolutionise medicine through the ability to simply switch certain genes off or on without altering the actual DNA sequence. It is vastly cheaper and easier to use than any previous gene editing techniques, which has opened up a whole host of new opportunities for research and experimentation.

The Key Laboratory of Regenerative Biology at the Guangzham Institute of Biomedicine and Health has already used this tool with dogs to create a genetically modified Beagle. Researchers knocked out both copies of a particular gene that produces a muscle-inhibiting protein to create a Beagle with more muscle and stronger running ability.

With gene editing, evolution is no longer a gradual process subject to natural selection but is immediate and controlled by humans. Chinese genomic institute BGI has already used gene editing to create micropigs the size of a Welsh Corgi and there are plans to start selling these new 'Bama' pigs as pets.

Since dogs were domesticated, breeders through the ages have bred selectively to create animals with certain desirable traits. What has previously taken generations to achieve can now be done quickly in a laboratory.

The health of our favourite dog breeds is likely to be improved in the future by gene editing as breeders can keep all the desirable traits they've always strived for without having to contend with the genetic disorders associated with them.

The University of California-Davis School of Veterinary Medicine is investigating a variety of genetic disorders in pets with a view to eradicating them. In one research project, a group has identified a gene mutation that causes Dalmatians to suffer from bladder stones and excessive uric acid levels. The findings have enabled a screening test to be developed so that affected dogs are not bred from in the future. With this chain of inheritance severed the condition will die out in the breed. Another group at U C-Davis has discovered a genetic mutation in Persian cats that causes polycystic kidney disease. As a result they have been able to develop a diagnostic test that can screen for the gene in very young kittens.

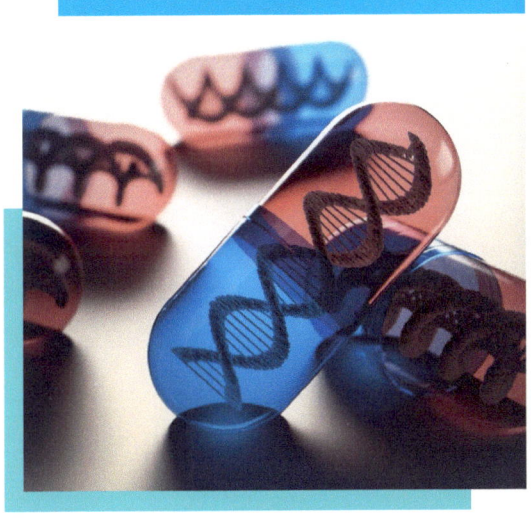

Gene therapy

Back in 2001 a group of scientists from the University of Pennsylvania and Cornell University College of Veterinary

Medicine used gene therapy to give sight to three puppies that had been born blind due to a genetic disease called LCA – a condition that also affects around 10,000 people.

Researchers identified the mutant gene that causes blindness and replaced it with a gene they had created in the laboratory by planting it inside a virus and injecting it into the puppies' eyes. The virus 'infected' the eyes with the replacement gene and within a month the dogs could see for the first time in their lives. The procedure is now being used to give sight to children who have been born with the same genetic disease.

In 2016 a team of researchers from the University of North Carolina published the findings from its study into dogs with FV11 deficiency – a genetic bleeding disorder where there is inadequate production of a blood-clotting protein.

Using a single gene-therapy injection with the missing gene enclosed inside the common cold virus, the condition was corrected. As blood, kidney and liver function tests all showed that the therapy was safe with no unwanted immune response, the next step will be to conduct clinical trials in humans with the condition.

Synthetic Biology

Biotechnology futurist, Andrew Hessel, believes that genetics has enabled biology to become programmable and he is approaching the treatment of cancer in this way by actually writing DNA. Hessel has created new genetic programmes that can be turned into programmable viruses to re-engineer cells.

Using a DNA synthesiser – a kind of chemical printer for DNA – he has created 'cancer killing' viruses that infect and kill cancer cells without harming the healthy cells around them. The beauty of these printable oncolytic viruses is that they can be produced individually and can be personalised.

As cancer is caused by a patient's own cells infecting their body, a 'one size fits all' approach to treatment is not the most effective.

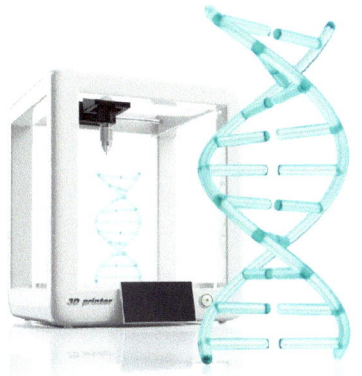

Currently the process of developing drugs and getting them approved is so time-consuming and costly that they need to be aimed at a huge number of people.

When everything is digital the cost drops dramatically and with DNA synthesisers you have the ability to print one offs without the need for a large pharmaceutical company.

Hessel designs a virus based on a patient's individual tumour and once printed plans an experiment to test it, which a computerised laboratory carries out.

He sends specific bacteria samples that the lab grows and once grown his virus is attached to the bacteria to see how it responds. Hessel has now teamed up with Dr. Bruce Smith from Auburn University College of Veterinary Medicine, who has been working with dogs with a common type of bone cancer, to trial a personalised approach with his printed oncolytic viruses.

EVERYONE'S AN ECO-SYSTEM

"It's time to lay rest the notion that germs jump into people and cause disease."

– Dr. Emanuel Cheraskin, physician and author

Your microbiome is the sum total of all the genetic material in your body, including all the microbes (non-human cells) like bacteria. There are about the same number of these micro-organisms in the human body as cells, roughly 30 trillion, and these influential microbes have, up until recently, been largely ignored by medical research.

Throughout the 20th century a number of micro-organisms were identified in the human body, isolated from the nose, mouth, skin and gastrointestinal tract. However, it wasn't until 2007 when the Human Microbiome Project (HMP) was launched that we began to fully understand the importance of microbial diversity. The HMP used the stool and tissue samples from more than 200 healthy volunteers to map the diversity of microbes in determining what a normal profile would be.

Just like the rainforest, the healthy human microbiome is actually a carefully balanced eco-system that controls essential functions in the body such as digestion and synthesising vitamins. Ensuring balance in the microbiome is essential to good health and the inclusion of microbiome mapping to modern medicine will radically change the way we view and treat chronic illness.

The University of Minnesota has set out to do for dogs what the HMP did for people with its Canine Microbiome Project. The project will comprehensively characterise the skin, oral and gut microbes in samples from 40 dogs over the course of two years to identify the core bacteria that are indicative of good health.

Researchers have found that many inflammatory and autoimmune diseases are linked to having either too many or too few microbes in the body. For example, it is now known that humans and dogs suffering from chronic inflammatory bowel disease (IBD) have significantly lower bacterial diversity than those who don't.

Penn Vet's Center for Host-Microbial Interactions is leveraging cutting-edge technologies to study microbes and disease with a view to gaining insights into how bacteria, parasites, viruses and other organisms interact with their human or animal hosts. One project is looking specifically at IBD in dogs, which is a condition characterised by inflammation and disruption of normal gut function.

In humans and dogs with IBD it has been noted that there is variation in how patients, and even the same patient over time, respond to treatment. It is believed that the 'good' bacteria living in the gut may not only influence the course of IBD but also the body's response to treatment.

Information from this study could lead to the development of a diagnostic test that profiles good gut bacteria in sick animals to guide treatment options. A study published in 2016 by the University of California San Diego Health Sciences documents the discovery of a pattern of microbes indicative of IBD. The differences were significant enough that researchers could distinguish between dogs with IBD and without IBD with more than 90% accuracy. Also in this study it was found that the gut microbiomes of dogs and humans were not similar enough to use dogs as a model when looking at human IBD.

Diet could play a big part in making changes to the microbiome of the gut as discovered by researchers at the University of Illinois. They found that introducing potato fibre to a dog's diet increased certain bacteria known to decrease IBD in humans and others known to promote good gut health.

Other studies are looking for ways to modify gut bacteria to help treat diseases such as cancer, immune diseases and obesity. By gaining an understanding of a normal canine microbiome vets can identify abnormalities and better provide guidance on probiotics and prebiotics.

As well as physical illness the microbiome has also been linked to mental health conditions including depression, anxiety and autism. A study carried out at UCLA looked at how the mood is affected by the microbiome by changing it with a pro-biotic.

Over four weeks one group of people ate a cup of commercially available pro-biotic yoghurt twice a day while another group was given a non-probiotic dairy product. Before and after the four weeks the participants were given functional MRI brain scans to monitor their brain activity during which they were shown images of human faces expressing emotions such as anger or fear. The brain scans showed that these emotional triggers affected the people in the control group more than the pro-biotic group, which was calmer and less stressed by comparison.

Research using the microbiome is still a relatively new field and we've only just begun to scratch the surface of what it could tell us about ours, and our pets', bodies. Studies are also branching out into other species, such as cats.

When a group of cat-loving biologists had their request for funding a feline microbiome project turned down they took to crowd-funding website Kickstarter to get it off the ground. The team has raised over $23,000 for its Kitty Biome project to map the entire feline microbiome.

This is the first research project to look at the feline microbiome in fine detail and it is hoped that the findings will help to better understand and treat medical conditions in cats, including diabetes and IBD. Backers of the project who want to discover the world of microbes living in their pet's gut have pledged $99 and sent fecal samples to the Kitty Biome lab for analysis.

In addition to a cat or dog's microbiome being a balanced eco-system it also affects and is affected by other microbiomes it encounters, including ours. Bacteria from a dog's coat and paws are easily transferred to the skin of the humans living with them. A study in 2003 at the University of Colorado showed that adults have more microbes in common with their own dogs than they do with dogs owned by others.

The study also showed that having a dog has an impact on the sharing of microbes between people living together - couples who own dogs had more bacteria in common than couples that didn't.

UCSF scientists also studying microbes have suggested that children growing up with dogs are less likely to develop asthma and allergies due to them training their developing immune systems with harmless bacteria.

A study being carried out at the University of Arizona is looking into whether dogs can directly improve the health of older people through the sharing of microbes. Over 50s who have never owned a dog before, or not had one for a long while, have adopted shelter dogs to test if good bacteria from the animals will transfer to them and improve their health.

BLEEDING-EDGE TECH

"10 years from now you'll prefer the robotic surgeon over its human counterpart."
– Peter Diamandis, founder of the XPRIZE Foundation that encourages technological development to benefit mankind.

Cutting-edge is old news - it's at the bleeding-edge of technology that the very newest and most exciting developments in healthcare can be found. Much of it sounds like it has come straight from the pages of a sci-fi novel but this is the foreseeable future for human and veterinary medicine.

You will soon find an AI-enabled robot physician assessing your health, or even performing surgery on you, and any replacement body parts you might require will be simply printed to order.

Robotic surgery

Already surgical robots are being used in human medicine as they enable surgery to be performed with precision, miniaturisation, small incisions, decreased blood loss, less pain and quicker healing time.

The Robotic-assisted Da Vinci Surgical System has been designed to help with complex surgery using a minimally invasive approach to operating on human cancer patients.

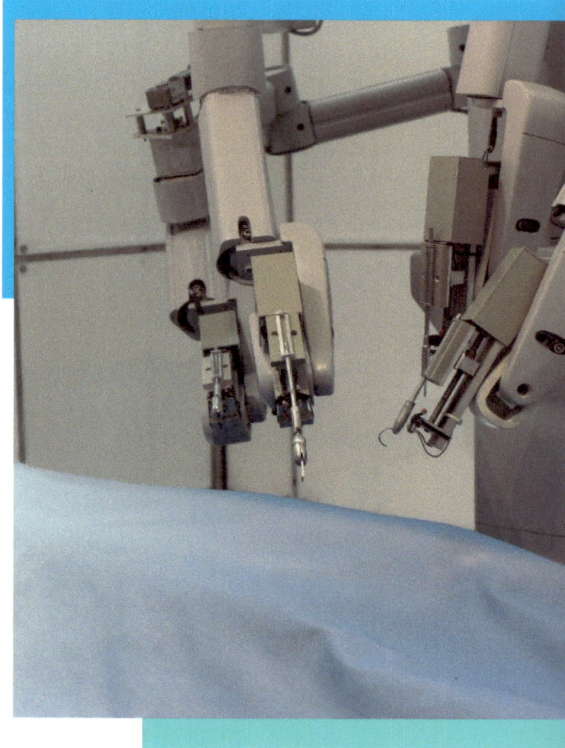

Small incisions are used to insert miniaturised instruments and a high-definition 3D camera, which the surgeon controls from a console.

This enables procedures to be carried out that would not be possible with a surgeon's hands.

Since 2000, the da Vinci Robot has been used in

more than 3 million procedures worldwide most commonly for hysterectomies and prostate removals.

In April 2015 a lion from Italy's Parco Safari delle Langhe Safari Park was the first animal to benefit from robotic surgery. The lion successfully had an adrenal tumour removed by a surgical robot at the Veterinary Hospital of the University of Lodi. The surgeon was able to work remotely from the robot's console with a 3D high-definition view of the animal's abdomen. This pioneering operation avoided the major tissue damage and the long, stressful recovery that traditional surgery would have meant.

In 2015 Verily Life Sciences (formerly Google Life Sciences) joined forces with pharmaceutical giant Johnson & Johnson to form Verb Surgical to further explore the possibilities of robotic surgery. Verb Surgical says it wants to combine the best medical device technology with cutting-edge robotics, imaging and data analytics to take things a step further than the da Vinci Robot. It's goal is to use algorithms to analyse on-screen images and do things like highlight blood vessels or display critical information for the surgeon on screen.

The aim of this would be to help surgeons see more clearly during surgery and to be able to easily access vital information needed during surgery. It may also be able to suggest the best places to make an incision for example based on analysis of a patient's individual medical history.

The next stage for robotic surgery would be to remove the surgeon altogether. In May 2016 a robot successfully performed soft tissue surgery all by itself for the first time, which is leading the way to autonomous operations. The STAR, or Smart Tissue Autonomous Robot, carefully scans soft, sticky intestinal tissue and combines these images with a 3D tracking algorithm to delicately stitch it up with a precision unmatched by any surgeon. By taking human intervention out of the equation it is believed that any complications could be reduced and the safety and efficacy of operations improved.

3D printing

3D printed prosthetics have been available to people with limb loss for the past few years. The ability to produce tailor-made, one offs at affordable prices has revolutionised the way in which prosthetics are delivered. In the US alone close to 200,000 amputations are performed each year yet prosthetics were previously considered an expensive luxury. A prosthetic has an average lifespan of five years, but growing children need frequent replacements and with traditional prosthetics costing between $5,000 and $50,000 – not to mention a wait of weeks or even months for it to be made – a new approach was clearly needed.

3D printing, also known as additive manufacturing, is a process in which a 3D object is produced under computer control made up of layers of material. It turns a whole object into thousands

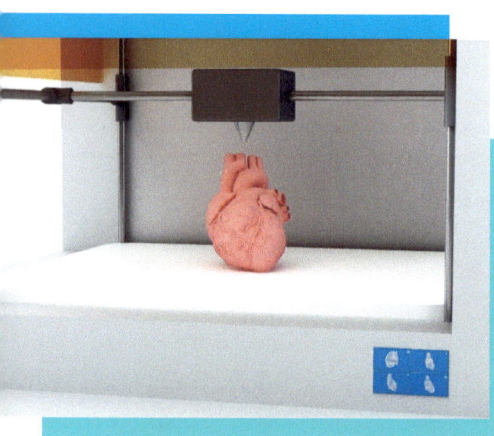

of slices and then constructs those slices from the bottom up to produce a 3D object. The beauty of this sort of production method is that a completely personalised prosthetic can be designed on a computer to fit its wearer exactly and produced much more quickly and cheaply than a traditional prosthetic.

A husky-cross called Derby is the first dog to be fitted with a pair of bespoke 3D printed legs. Derby was born with small, deformed front legs and severely struggled with his mobility, even with a set of doggie wheels. Fortunately Derby's foster carer worked for a 3D printing company that was happy to take on the challenge of creating the perfect pair of legs for him. After several modifications to the original design Derby's final pair of legs enable him to run, jump and even sit just like a normal dog.

The same company that created the legs for Derby has gone on to help more than 10,000 dogs with damaged cruciate ligaments by 3D printing special implants for them. The 3D printed metal orthopedic knee implants have meant that surgery for the condition is now easier and patients can benefit from a faster recovery time compared to previous methods.

The next step for 3D printing is creating human organs for transplant so that organ donors are no longer required. California-based company Organovo is already making progress towards this by printing liver tissue and believes in the next few years 3D printed tissues could be used in human medicine.

3D printing organs, which comprise different types of living cells in need of nourishment, presents a much greater challenge. The process involves scientists growing cells from biopsies or stem cells and then arranging them three-dimensionally in the way they would be found in the body. When the cells have been printed they should begin to signal each other, fuse and organise themselves into a collective system.

A group of scientists at Harvard have succeeded in using a 3D printer to make human tissue that actually includes rudimentary blood vessels. This is the basis for their plan to create a fully functioning kidney. There is also a team at the University of Louisville that is working on a 'biofical' (a blend of natural and artificial) heart using a 3D printing technique. The ultimate goal being to put all the pieces of the heart together using a patient's own cells to create a viable working heart for transplant.

3D printing is also being used to make medical equipment and 3D4MD is a company that is building an iTunes style library of downloadable designs to print, such as medical instruments. One use they have found for this is on the International Space Station where medical devices can be printed onboard only when they are needed. Even places without electricity can benefit from this service as it is possible to buy a 3D printer the size of a small suitcase that runs on solar-power.

In 2015 the US Food and Drug Administration approved the first 3D printed pill. Spritam Levetiracetam is an epilepsy drug and its manufacturers used 3D printing along with a new technique it had developed to make a more porous pill that dissolves quicker.

For 50 years tablets have been manufactured in factories in huge volumes before being shipped around the world. With the advent of 3D printing drugscan be made in much smaller amounts on demand and the compositions can actually be tweaked to an individual's requirements.

Researchers at UCL School of Pharmacy have been using 3D printing methods to produce tablets in a variety of shapes, from pyramids to doughnuts, to alter the release rate of the drugs. These findings will help manufacturers design pill shapes that can accurately release the drugs at precisely defined rates. In the future you, or your pet, could just take one pill a day containing all your medications for that day, completely customised to your unique requirements and be released at the right dose at the right time.

Virtual Reality and Augmented Reality

Imagine being fully absorbed watching a medical procedure or surgery where you can pause the action, walk 360 degrees around the patient, zoom in or out, slow down or repeat sections again and again. This is the future of veterinary training when virtual and augmented reality is involved.

Virtual reality (VR) is the generation of realistic images and sounds to simulate a user's presence in an unreal setting. This could be a replication of a real environment or an imaginary setting. The user normally wears a headset or goggles to immerse themselves in this other reality in which they can look around and interact with certain features.

Virtual reality technology can enable trainee surgeons to gain valuable experience in a safe environment where they can learn skills without any risk to patients. This style of training builds confidence in surgeons because performing a surgical procedure on a virtual patient allows them the freedom to make mistakes.

Virtual reality replaces the real world with a simulated one whereas augmented reality gives you a real-word view that is enhanced by computer-generated video, graphics or data. You could have menus pop up in your field of vision giving you important information that is relevant to that time or situation.

A vet, while his or her hands are busy, could instantly access a patient's details right in front of them. In 2016 the world's first virtual reality operation was broadcast live on the internet.

Surgeon Shafi Ahmed performed a three-hour operation to remove cancerous tissue from the bowel of a patient while medical students, trainee surgeons and interested members of the public watched from their armchairs.

Virtual reality technology has the advantage over straight video for observing operations as viewers are able to focus not just on

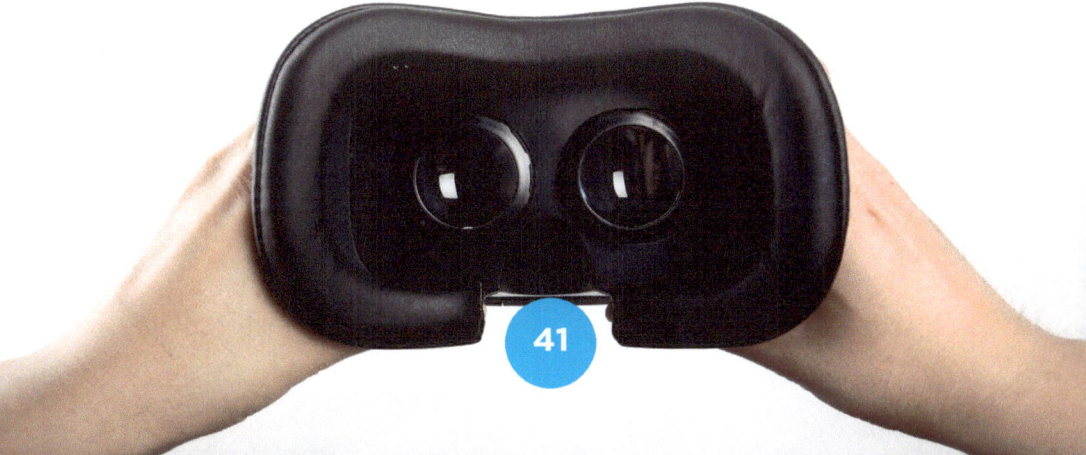

what the surgeon is doing but also the other members of the team at any time. Ahmed, who co-founded the healthcare company Medical Realities, said he believes VR and AR could play an important role in training medical students.

Artificial Intelligence

Machines with artificial intelligence (AI) have the ability to mimic the cognitive functions of a human brain, such as learning and problem solving. In the 21st century AI techniques became an important part of the technology industry and are now used to help solve many of the challenges faced in computer science.

Super computers equipped with artificial intelligence are able to analyse enormous amounts of data and cross-reference it quickly in a way no human ever could. This has opened up a whole new way to diagnose disease and is already proving to be very successful.

IBM's super computer, Watson, correctly diagnosed a patient within minutes in 2016 - something doctors had failed to do after months.

The woman from Japan was being treated for leukemia but doctors were puzzled as to why the treatment she was being given wasn't being effective. After just 10 minutes studying the patient's medical information Watson was able to cross-reference her condition against 20 million oncological records that had been uploaded to it by doctors from the University of Tokyo's Institute of Medical Science. Watson found that the patient had a different form of leukemia to the one that had previously been diagnosed, which required different treatment.

Watson uses cognitive computing, which means it 'thinks' more naturally, and can provide answers from large amounts of data. It can be fed millions of documents on a subject and uses its machine learning capabilities to identify questions and work out the most logical answer. This question answering capability enabled Watson to win the American gameshow Jeopardy in 2011 against two human contestants. Watson did this by analysing 200 million pages of content including the full text of Wikipedia.

A similar super computer has been developed specially for the veterinary profession. Sofie is powered by Watson's cognitive computing algorithms and has been trained by veterinarians for veterinarians.

Vets can simply type a question into the "Ask Sofie" app on their smartphones based on a patient's initial consultation and the computer returns focused, evidence-based treatment options. Sofie works in a different way to a search engine as it actually understands natural human language rather than relying on keywords.Scientists at the UCLA NanoSystems Institute have used AI to develop a new technique for identifying cancer cells in blood samples faster and more accurately than the current methods. The AI microscope is able to analyse 36 million images a second looking for cancer cells and is finding the cells with 95% accuracy.

MONITORING AND EARLY WARNING SYSTEMS

We need to get upstream. Right now we're treating people at the end. When things have already gone wrong."
– Tony Young, NHS England

In the veterinary medicine of the future, owners will play a greater part in their pets' wellbeing by having the ability to pick up on the early warning signs of illness. For example, the technology exists, through a company called Petnostics, to enable owners to test their pet's urine at home. Petnostics has developed a cup with an integrated urine test strip, which owners can use to collect a sample and test it.

Once the test strip has changed colour an owner can take a

photograph of it with their smartphone and the Petnostic app will alert them to any potential health issues. The test strip on the Petnostics cup is the same as the ones used by vets in clinics and can indicate many conditions including diabetes, urinary tract infections, bacterial infections, liver disease and dehydration.

The smartphone is increasingly becoming an essential tool in the first aid kit and there are lots of companies leveraging apps to help with health monitoring and diagnosis. These technologies are likely be finding their way into veterinary medicine too. A company called Cordio Medical is using clever multi-layer voice analysis technology to pick up on the changes in a person's tone of voice that can indicate Congestive Heart Failure.

Behind a simple smartphone app is a sophisticated cluster of specially developed algorithms that analyse the voice to detect an early build-up of fluids in the lungs, before the appearance of any physical symptoms. Alerts are generated once a threshold amount is crossed so that the phone user knows to seek medical advice. Could we be facing a future in which our dogs will be barking into smartphones to detect heart failure? It certainly seems like anything is possible.

The idea of a hand-held medical monitor is not a new one and fans of the TV show Star Trek will already be familiar with the tricorder. This fictional multifunctional medical device was used on the Starship Enterprise to diagnose disease and collect bodily information about a patient. The idea of having one device that can monitor and diagnose a patient appealed to those at the XPRIZE Foundation (an organisation that runs competitions to encourage innovation through technology) and it launched its Tricorder XPRIZE contest in 2011.

The competition brief is for competing teams to design and produce a working prototype tricorder device that is capable of capturing health data and diagnosing a set of 12 different diseases. This could include blood pressure, respiratory rate and temperature, and would be combined with artificial intelligence to make sense of the information and identify trends. The innovations that come out of this competition are likely to shape the future of human, as well as veterinary, medicine.

The University of California San Diego has already developed the first flexible, wearable tricorder device capable of monitoring both biochemical and electric signals in the human body. The Chem-Phys patch is worn on the chest where it simultaneously measures a whole suite of signals in the body throughout the day, which it then communicates wirelessly to a smartphone, smart watch or tablet. The patch is primarily being aimed at athletes looking to improve their performance and doctors monitoring patients with heart disease, but there are almost endless applications for this style of monitoring.

Verily Life Sciences has teamed up with pharmaceutical company Novartis to develop a smart contact lens that continually monitors glucose levels in a person's tears. The smart lens connects wirelessly to the wearer's smartphone to provide a read out of data, or alert the wearer to any sudden changes in their blood sugar level. This would revolutionise the way in which people with diabetes check their levels and would remove the need for frequent finger prick tests. It is easy to see how this technology could be adapted for use with diabetic pets in the future. Currently between 1 in 100 and 1 in 500 dogs and cats develop the disease.

Tufts University in Massachusetts is developing 'smart thread' which will turn surgical sutures into sensors that are actually sewn into the body. This would allow a more detailed overview of the body than any skin-mounted sensors currently in use. The advantages of using thread as a sensor are that it's cheap, flexible, and doctors and vets already have plenty of experience stitching up patients. Covering thread with conductive ink creates electrodes for recording mechanical or chemical activity.

By running a small electric current through the suture the forces surrounding it can be measured to monitor healing. There are lots of potential applications for smart sutures, including use with diabetic patients who must have wounds closely monitored as failure to heal can lead to amputation.

A group of engineers from UC Berkeley are harnessing the information in sweat as an effective way of monitoring the body. The flexible sensor system it has designed measures electrolytes in a person's sweat, calibrates the data based on skin temperature and sends real-time results to a mobile device, alerting users to health problems such as fatigue, dehydration, and dangerously high body temperature.

The prototype smart wristband and headband incorporate five different sensors that measure skin temperature, glucose, lac-tate, sodium and potassium in sweat.

There is a new way of monitoring medication in patients with chronic illness that involves an ingestible sensor activated by stomach acid. Proteus Discover combines a tiny sensor with a small wearable sensor patch and app that can run on a smartphone or tablet. The sensor, which is about the size of a grain of sand, is implanted in a patient's normal medication and activated once swallowed.

When the sensor reaches the stomach it transmits a signal to the patch to tell it that the medication has been taken and a digital record is sent from the patch to the patient's mobile device along with data such as heart rate and activity. Patients can more easily monitor their medication use and, with their permission, the information can be accessed by healthcare providers and caregivers to improve service.

This emerging field, nicknamed Smart Medicine, will allow medical providers to assess, track and monitor patients in between clinic visits, allowing more personalised and interactive care for people with chronic medical conditions. Using Smart Medicine with animals would also be a valuable tool for the rising number of pets suffering from chronic illness. Owners could easily keep track of the medication their pet is taking, especially useful for those animals who manage to spit their pills out in secret long after their owners think they've swallowed them!

Verily Life Sciences has big ambitions when it comes to human healthcare, especially in the field of monitoring. Ultimately it wants to introduce systems that continuously monitor the body, comparing the data it receives to a baseline study of a healthy human. The aim is to be able to detect something wrong in the body before a person even begins to feel unwell.

Using nanotechnology, Verily wants to create wearable and ingestible technology that can detect all sorts of things going on in the body. The first thing you know about an illness could be an alert from your smartphone telling you to see your doctor.

This constant monitoring will change veterinary medicine as we know it. No longer would owners identify symptoms of ill health in their pets and take them to their vet for investigation. Illness could actually be a thing of the past as an early warning system could flag up issues before you even know something is wrong.

Animals can't tell us when they are feeling unwell and their survival instincts actually make them hide any signs of illness as best they can. This would no longer be an issue with constant monitoring and the earlier that disease is identified the more chance there is of complete recovery.

Dr. Daniel Kraft, Founder of Exponential Medicine, says that we are moving from a 'sickcare' mindset to a 'healthcare' one and that continuous monitoring in the future will act in a similar way to the systems in modern cars. In a talk he gave at the Exponential Medicine conference in 2015 he said, "Our modern cars have three or four hundred sensors in them but you don't care about any one sensor, you care about when your 'check engine' light goes on."

In the future people will be alerted to problems in their bodies (or their children's or pets') by a 'check body' light. This will prompt you to make a doctor or vet appointment to fix the issue before it begins to make you or your loved ones feel unwell.

Constant monitoring has the potential to improve quality of life for all.

GOING GLOBAL

"Crowdsourcing is the ultimate disruptor of distribution because in a most zen-like fashion, the content is controlled by everyone and no one at the same time."
– Jay Samit, digital media innovator

The Earth appears to be a much smaller planet today as we've all become connected with the ability to communicate in real-time, face-to-face with other people all over the world. In the field of veterinary science leading experts from different countries can work together easily, comparing research and sharing their findings to further medicine for the benefit of all.

Telemedicine

Telemedicine is the use of telecommunication and information technology to provide medical advice and expertise from a distance. In this connected world the best specialists from around the globe can be consulted easily at the touch of a button.

Telemedicine has been growing in popularity in human healthcare with doctors consulting with their patients by phone, email and Skype for added convenience, and this has led to a reduction in hospital readmissions.

In the veterinary world, however, the primary diagnostic is still physical examination. Any translation by an owner acting as an intermediary for their pet is not likely to give a clear and accurate picture. For telemedicine to be an effective tool for vets it would need to be used as part of a more holistic approach and not on its own as a replacement for a consultation in clinic.

Nonetheless, some owners with pets living in remote places have difficulty in reaching a vet and can only do so in an emergency situation. What if an owner just wanted some advice or to find out if a visit to the clinic is really necessary?

I-Vet is an online veterinary service that utilises video conferencing to provide access to veterinary care for those living in remote areas or isolated by disability. I-Vet is based in the Northern Territory of Australia, where it can mean a 10-hour drive to the nearest vet.

The I-Vet website gives Skype-style video consultations with an emergency on-call vet 24 hours a day and can also provide instructional videos for simple treatments. I-Vet customers can even have vaccinations couriered to them. The vaccines come with clear instructions and if an I-Vet vet witnesses owners administering them via video conferencing they will be happy to sign a vaccination certificate.

It would be helpful for a vet who is consulting remotely to be armed with as much detail as possible about a pet's vital signs and there are ways that this might be possible in the future.

Remote healthcare monitoring is something that is likely to be adopted in human medicine in the near future following successful pilot tests. The E-Care project was developed to capture, transmit and distribute health data to doctors and carers. It is designed for patients with chronic or long-term conditions that need to be closely monitored but might not be able to make regular clinic visits.

The E-Care remote monitoring system consists of a series of monitors that measure activity, temperature, pulse, blood pressure and glucose. A central system takes this data and compares it to a patient's medical record. If there is a big change in any of the readings it will send an alert.

The module allows a patient or carer to connect directly with a doctor or specialist via audio or video conferencing. A system where a pet owner could use monitors to measure their cat's or dog's vital signs could be integrated into an online veterinary service.

The Cellscope is an iPhone attachment with an otoscope on it that is now commercially available and being used by parents of children with recurrent ear infections. An app on the phone guides the user to take photos and videos of the ear canal, which can then be sent to a specialist for a real-time remote consultation within two hours.

It has proved to be not only a very useful diagnostic tool that reduces the number of visits to the doctor but a good way to follow up treatment and monitor a child's condition. There is also a stethoscope available that fits onto a smartphone and similar things could be developed for other parts of the body too, which could easily be used with pets.

The Power of Crowds

Crowdsourcing is a way of harnessing information from a big group of people via the internet to help with a particular task or project. In 2015 an American vet student realised the power of crowdsourcing to match ill pets with clinical trials, therefore sparing the lives of laboratory animals.

The One Health Company recruits pet dogs and cats for trials of new treatments for diseases that are similar in animals and people.

Currently healthy animals are given diseases for the purposes of clinical trials and are euthanised at the end.

This new crowdsourcing model enables the owners of cats and dogs with medical conditions such as diabetes, cancer and heart disease to enroll their pets via the One Health Company website where they will be matched to any upcoming clinical trials. Any treatments that are successful in trials will then be used in veterinary practice as well as human medicine.

There are currently more than 450,000 pets on the One Health Company's website and so far they have been matched to clinical trials in ocular melanoma, transitional cell carcinoma and irritable bowel syndrome.

Crowdsourcing data has been a valuable tool in enabling artificial intelligence. The Sofie app that vets can ask for advice is powered by IBM Watson's artificial intelligence with access to a vast amount of patient data that has been collected via crowdsourcing from a number of veterinary hospitals.

Crowds are already being used to help diagnose rare human illness and US-based startup CrowdMed has developed a web tool that does just that. According to CrowdMed, "no single individual, even a doctor, can keep track of thousands of unique medical disorders, with hundreds of new ones discovered each year". Instead of relying on individual doctors, CrowdMed utilises the collective intelligence of hundreds and is finding it produces astonishingly accurate diagnostic suggestions in just hours.

In the past, vet practices have kept their own patient's medical records, firstly on paper and then on computer systems.

There are a whole host of practice management software programmes used in vet clinics, which provide access to medical records, manage finances and help grow the businesses but these are all centered on individual practices or groups.

VetCompass is a UK-based system, which is a collaboration between the Royal Veterinary College and the University of Sydney. It collects data from around 450 veterinary practices about every aspect of pet care. The data is then analysed and trends identified. The data can show, amongst other things, how long cats and dogs are living and what's likely to kill them. Data also reveals which treatments are most efficient.

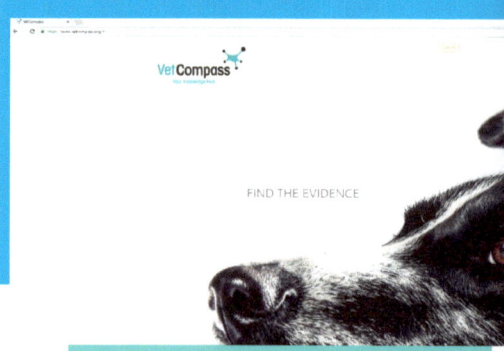

Prior to the VetCompass system any comparisons between practices were anecdotal. Internet-enabled collating of data will result in better medicine with the ability to understand trends and be more able to take preventative steps.

Imagine a system that all the veterinary practices in the world were connected to and what opportunities that crowdsourcing of data could have on the future of our pets' health.

Crowdfunding is the crowdsourcing of funds to pay for things that are in the public interest and there are a number of websites that run funding appeals for projects. The beauty of crowdfunding is that people contribute financially to the projects that they see as having a direct influence on themselves.

Examples in the pet world are the Tailio smart cat litter tray and the Kittybiome feline microbiome project that both used Kickstarter to raise enough funds to get their projects off the ground. The ability for individuals to contribute financially to projects such as this opens up areas that were previously not possible without large sponsors or company backing.

Crowdfunding gives power to the people and power to pet owners when it comes to the next steps in companion animal health.

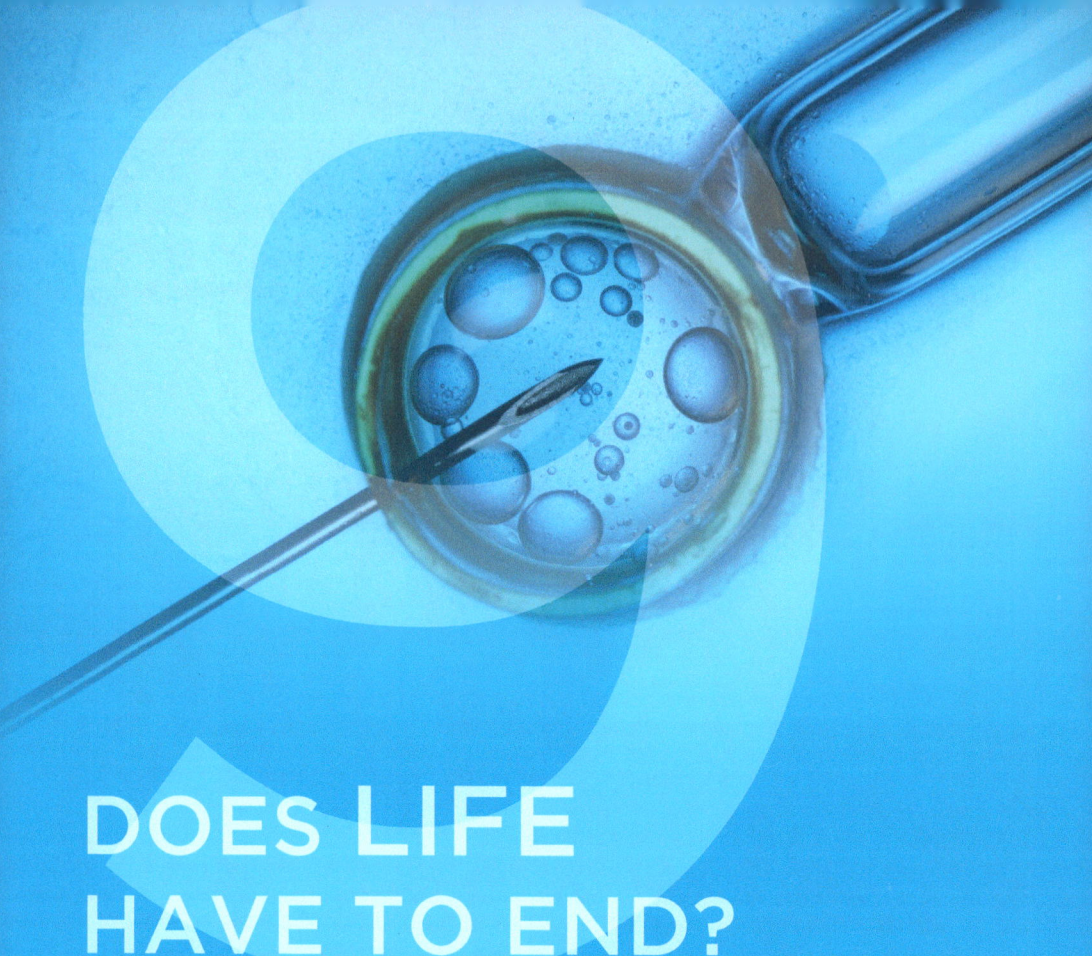

DOES LIFE HAVE TO END?

"I laughed when Steven Spielberg said that cloning extinct animals was inevitable. But I'm not laughing anymore, at least about mammoths. This is going to happen. It's just a matter of working out the details."

– Hendrik Poinar, evolutionary biologist

It's a topic no pet owner wants to think about but at some point you may be faced with the difficult choice between euthanasia or pursuing treatment options for your pet. One of the reasons why this is such a tricky decision is the difficulty in which we find assessing an animal's quality of life. This could be made easier thanks to a quality of life survey that has been developed by researchers at Michigan State University.

Researchers looked at canine cancer patients with the aim of helping owners decide whether to prolong their pets' lives with additional medical procedures or not. The dogs' owners were given questionnaires to complete at the time of diagnosis about their pets' behaviour and then follow-up questionnaires were given three and six weeks later to document changes during chemotherapy treatment. The dogs' vets were also given questionnaires to complete based on their own observations.

Veterinarians are able to use their training, experience and scientific knowledge to objectively assess an animal's quality of life but at home owners rely on their own subjective impressions of their pet's wellbeing. Researchers aimed to isolate the emotions in the survey to enable people to make the best decisions for their pet and for themselves.

The findings from the pet owners and vets were closely matched, particularly in the areas of play behaviour, clinical signs of disease and canine happiness as perceived by the owner. This told researchers that the questionnaire was a helpful way to find common ground for treatment decisions.

It was concluded that the survey makes an effective indicator of quality of life that can be used in animal cancer clinics and could be a valuable tool for pet owners who may not know the changes they should be looking out for that indicate quality of life. It should therefore help decision-making and enable owners to assess their own pet's wellbeing before making difficult choices about their future.It is now hoped that the survey could be adapted for use with animals that are suffering from illnesses other than cancer.

Cloning

Inevitably at some point we must say a final goodbye to our beloved pets and owners cope with this dog or cat-shaped hole in their lives in different ways. Some people find it helpful to get a new pet as soon as possible in an effort to fill the gap but, no matter how hard you try to find a replacement it will never be the same. However, if you have a spare $100,000 you can have your pet cloned and re-live their lives as puppies or kittens.

The first successfully cloned dog, an Afghan Hound named Snuppy was born in 2005, nine years after the birth of Dolly the sheep. Snuppy was cloned by a team from Seoul National University whose leader Hwang Woo-suk went on to form his own cloning company – Sooam Biotech – and has since gone on to offer commercial
pet cloning. It was Sooam Biotech that produced the world's first commercial dog clone in 2011. The dog cloned was a Labrador called Lancelot owned by Edgar and Nina Otto from Florida. The couple paid $155,000 for the pup clone, who they named Lancelot Encore, and they were very
happy with the outcome saying that they bonded with Encore straight away as he looked and acted just like the original Lancelot.

The clone of an animal shares the exact genetic identity as the original, just as in naturally occurring identical twins. The complex cloning process involves DNA being taken from an animal while it is still alive, or shortly after death, and implanting it into a 'blank' donor egg that has had its nucleus removed. The egg is then given electric shocks to trigger cell division before being implanted into a surrogate mother. Sooam Biotech has produced more than 700 clones of dogs for commercial customers and in 2015 announced it was teaming up with Chinese biotech company BoyaLife Group to open the Tianjin Animal Cloning Factory. It will be the largest cloning facility in the world and aims to produce up to a million cattle embryos a year to meet the demand for quality beef in China.

In 2015 Sooam Biotech produced two clones of a deceased boxer dog for a British couple. This was a first for Britain and a first for Sooam Biotech as, when the cells were taken, the dog had already been dead 12 days. The longest delay previously for successful cloning had been five days so this opened the possibility for increasing the timescale in which samples would need to be collected from a dog post-mortem.

For owners who are not in a position to be able to pay $100,000 for a clone of their pet, Sooam Biotech offers a cell storage service. The company will cryopreserve (preserved by cooling to very low temperatures using liquid nitrogen) a pet's cells for $3,000, which can be defrosted at any time in the future to make a clone.

The first cat clone, named Carbon Copy or CC for short, was born in December 2001 at Texas A&M University.

Genetically CC was an exact copy of the original cat, Rainbow, just as if they were naturally occurring identical twins, yet they looked very different from each other. Rainbow was orange mixed with patches of black, with a white belly and legs while her clone was a tabby.

This is because the pattern of colours on multi-coloured animals is determined by events in the womb, rather than by genes. It is not know how long CC lived but she had a litter for four kittens in 2006 and reached at least 10 years old. Clones may be genetic copies but they can never be completely identical because there is so much that is dictated outside of the genes. An animals' character is influenced by early experiences and its environment so it is unlikely that a clone will have exactly the same character and traits as the one cloned.

The cloning of pets is a controversial topic that has divided opinion and there are currently no regulations governing pet cloning. It is, however, illegal to clone human beings and in 2015 the European Parliament voted to make the cloning of farm animals illegal.

Cryonics

If cloning seems like a risk without the guarantees of an identical looking replica you could always consider having your pet cryopreserved in the hope that future medicine might allow him to live again.

Cryonics is the practice of preserving bodies using very low temperatures until such time that future technology can restore them to full health.

It sounds like science fiction but there are already 1,000 people worldwide who have made legal and financial arrangements to be preserved in this way and some of these have, or plan to

have, their pets preserved too in the hope that they can be reunited one day.

President of the Cryonics Society of Canada and nurse Christine Gaspar cryopreserved her cat Marmalade in 2013.

Following euthanasia Marmalade's blood was drained and swapped for glycerol to prevent ice crystal formation in the cells, before being stored in a big flask of liquid nitrogen called a cryostat. Marmalade currently resides at the Cryonics Institute in Michigan where there are a total of 120 pets waiting to be restored. Cryonics advocates insist they are not dead, as real death occurs only when cell structure and chemistry become so disorganised that no technology could restore the original state. The object of cryonics is to prevent death by preserving sufficient cell structure and chemistry so that recovery remains possible by foreseeable technology.

Cryonics is based on modern science and predictions about the future of technology but providers are very open about the fact that it is very much an experiment and there are no guarantees future restoration will be possible.

PETS OF THE FUTURE

"We are possibly witnessing the dawn of a new era, the digital revolution with likely effects on pet ownership, similar to the industrial revolution which replaced animal power for petrol and electrical engines."

– Dr. Jean-Loup Rault, University of Melbourne's Animal Welfare Science Centre

Robotic pets

Over half of the people in western society spend their lives with pets and our shared history goes back tens of thousands of years. But will a technological revolution change this human-animal relationship?

Powered by artificial intelligence and made to resemble cats and dogs, the robot pets of today are able to mimic live animals more naturally than ever before. The increasing urbanisation of the planet is likely to see animal lovers turning to robotic pets in the future as more convenient and practical pets for the modern age.

Dr. Jean-Loup Rault predicts that real pets will become a luxury possession only owned by the small percentage of the population able to fulfil their needs in terms of space, social and mental stimulation. As a pet owner it might be hard to believe that a robot could occupy the place in your heart where your dog or cat currently resides but robot pets could become the norm in less than a decade.

Astrophysicist Neil de Grasse Tyson is quoted as saying, "Cars have replaced horses. Home security systems have replaced guard dogs. Internet kittens are cuter than your kittens.

Pooper-scooper laws are unnecessary for cuddly stuffed animals on your bed. And your pet bird should never have been kept in a cage to begin with. Robot pets are inevitable and possibly overdue."

The ability of humans to respond emotionally to robot pets has led to the use of robotic animals in hospitals, with the same therapeutic effects as pets used for animal therapy. The PARO robot seal pup displays emotional responses to external stimuli via a number of sensors on its body, and it is designed to have a positive psychological effect on the people who interact with it.

Through interaction PARO gradually learns to develop a personality that its owners like. There are 1,300 of these robot seal pups being used in Japan, particularly with dementia patients as an alternative to sedatives.

Sherry Turkle, anthropologist at MIT, conducted research into the ways children respond to smart toys like the Furby, Tamagotchi and Sony's Aibo robot dog.

The Tamagotchi digital pets were introduced in Japan in the 1990s and were originally designed for teenage girls to give them an idea about what it would be like to take care of children. All toys such as these require nurturing, which encourages children to take care of them and to care about them. Some children have said they prefer these digital or robotic pets to actual cats and dogs because they do not grow old and die.

Turkle is quoted as saying, "People used to buy pets to teach their children about life and death and loss. We are now teaching kids that real living creatures are risky, while robots are safe." She has studied people's feeling towards robots and has found that a culture shift has occurred over time. In the 1980s and 90s people she interviewed told her that love and friendship are connections that can only exist between humans. Now people often tell her that they believe robots could fulfil this role too. She concludes that humans are programmed to respond to creatures in a caring way, even if they are artificial.

Robotics company Boston Dynamics has produced some astounding robot dogs that walk just like the real thing on four legs, can run, climb stairs, balance on uneven terrain and even pick themselves up if they fall over.

Boston Dynamic's SpotMini robot, which was unveiled in 2016, weighs about the same as an average Golden Retriever and can assist with housework such as loading the dishwasher. It looks like the pet of the future will have the ability to fulfil an owner's emotional needs as well as becoming a household assistant. It will not require feeding, or cleaning up after, so will fit perfectly into people's busy lives and small homes.

Jason Silva, host of National Geographic's Brain Games programme believes that, rather than replace our pets with robots we'll continue to engineer them. He is quoted as saying, "Remember that dogs were our first biotechnology project. We bred them into existence in their current form. We engineered them."

Uplifted animals

Uplifting is a science fiction concept that can be traced back to H.G. Wells' novel The Island of Dr Moreau in 1896. The idea is that certain species of animal could have their intelligence 'uplifted' by humans using interventions such as genetic engineering. It was more than 70 years later when genetic engineering began being used as a way of directly manipulating DNA. Did H.G. Wells predict the future and could uplifted pets really become a reality?

Research into genetics and neurological disease has shown that it could be possible to enhance animals' intelligence by making modifications to their brains. Humans share most of their genes with all other mammals and biologically there is not a huge difference between them.

Therefore it should be possible to work out which genes are responsible for the minor differences, such as those that allow humans but not chimpanzees to talk..

In 2012, scientists demonstrated that a brain implant could improve the thinking ability in primates. Researchers from Wake Forest Baptist Medical Center at the University of Southern California, implanted an electrode array into the cerebral cortex of five rhesus monkeys that were under the influence of cocaine.

Researchers found they were able to restore, and even improve, the monkeys' decision-making abilities using the implant. Although an experiment such as this demonstrates the potential to increase thinking capabilities in animals, many bioethicists question whether it would be as straightforward as this.

Significantly increasing an animal's intelligence would require extreme modification in nearly every aspect of their bodies and minds. If we desire uplifted dogs to function as intelligent humans we would want to overcome the short lifespans that they have too. A pet with near human intelligence would have an even stronger bond with its owner and its death after only 12-15 years would be akin to losing a child.

Bioethicist George Dvorsky believes that when the technology becomes available to enhance the intelligence of animals we should do and says it would be negligent of humans to purposely withhold intelligence enhancing technology from them.

Sci-fi writer David Brin wrote a series of novels in the 1980s called Uplift where humans genetically uplifted the intelligence of dolphins, chimps, gorillas and dogs. His view is of a future where these wise animals would serve on our councils and offer their own insights, enriching an Earth civilisation that is no longer only human. Brin believes that the uplifting of dogs would benefit us all, merely continuing a process we have engaged in for at least the last 10,000 years. He says that by increasing dog intelligence and abilities it would give us fresh insights into intelligence itself, as well as helping dogs partner humans in even more meaningful ways.

The future of veterinary medicine is advancing at an exponential rate with brand new technologies being tested and amazing discoveries being made every week. Keep ahead of these exciting developments by signing up to futuristvet.com.

REFERENCES

INTRODUCTION

Lucien Engelen's fourth industrial revolution
slideshare.net/lucienengelen/xmed-2015-lucien-engelen/11-Emotionbased_articial_intelligence

Dr.Leroy Hood, P4 Medicine Institute
http://p4mi.org/leroy-hood-md-phd

In 1900, horses comprised 80% of a vet's workload. In 1910 this dropped to 10%. American Veterinary Medical Association fact sheet.
sites.si.edu/animalconnections/AVMA%20150th%20Fact%20Sheet%202014.pdf

NEW MEDICINE

The One Health Initiative
onehealthinitiative.com

University of Pennsylvania School of Veterinary Medicine
vet.upenn.edu

The Humanimal Trust
humanimaltrust.org

Functional Medicine
functionalmedicine.org

Banfield Pet Hospital state of health report
banfield.com/state-of-pet-health

The Precision Medicine Initiative
https://ghr.nlm.nih.gov/primer/precisionmedicine/initiative
whitehouse.gov/precision-medicine

National Center for Complementary and Integrative Health
https://nccih.nih.gov/

THE INTERNET OF ANIMALS

Smartphone statistics
statista.com/statistics/330695/number-of-smartphone-users-worldwide/

Internet of Things statistics
cisco.com/c/r/en/us/internet-of-everything-ioe/internet-of-things-iot/index.html

Tailio smart litter box
tailio.com

SmartFeeder from Petnet
petnet.io

Obedog ProBowl
obedog.com

PetBot
petbot.com

Wearable technology for pets on the rise
idtechex.com/research/articles/wearable-technology-for-animals-a-2-6bn-market-worth-watching-00006576.asp

Voyce health monitor
voyce.com

Whistle activity monitor and tracker
whistle.com

Melody Moore Jackson, What if dogs could talk? TED talk
youtube.com/watch?v=aMuVpnUEIMg

Dairy cow sensor technology
extension.umn.edu/agriculture/dairy/health-and-comfort/cow-sensor-technologies/index.html

UNZIPPING GENES

National Human Genome Research Institute
genome.gov/10001772

23andMe genome sequencing
23andme.com

Raymond McCauley on the DNA revolution
singularityhub.com/2016/10/09/3-dna-technologies-that-will-forever-change-your-home-life

Dog Genome Project
broadinstitute.org/scientific-community/science/projects/mammals-models/dog/dog-genome-links

Embark advanced dog DNA test
embarkvet.com

Embark and Cornell University dog genetics study
http://cornellsun.com/2016/03/08/cornell-vet-school-researchers-publish-largest-ever-study-of-dog-genetics

First gene-edited dogs reported in China
technologyreview.com/s/542616/first-gene-edited-dogs-reported-in-china

Gene-edited micropigs
nature.com/news/gene-edited-micropigs-to-be-sold-as-pets-at-chinese-institu-te-1.18448

University of California-Davis research project into Dalmatian bladder stones
vgl.ucdavis.edu/services/Hyperuricosuria.php

University of California-David research project into Persian cat Polycystic kidney disease
vgl.ucdavis.edu/services/pckd1.php

University of Pennsylvania and Cornell University restore sight to blind puppies
http://articles.orlandosentinel.com/2001-04-29/news/0104290149_1_blindness-gaines-ville-eye-disease

University of North Carolina fix blood-clotting disorder using gene therapy
hemophilia.org/Newsroom/Medical-News/Researchers-Make-Gene-Therapy-Break-through-in-Dogs-with-Factor-VII-Deficiency

Andrew Hessel hacking dog cancer
wired.co.uk/article/andrew-hessel-autodesk
andrewhessel.com

EVERYONE'S AN ECO-SYSTEM

Human Microbiome Project
commonfund.nih.gov/hmp/overview

Canine Microbiome Project
tc.umn.edu/~joh04207/JohnsonLab/Canine%20Microbiome.html

Penn Vet research into IBD
vet.upenn.edu/research/centers-initiatives/center/center-for-host-microbial-interactions

University of California San Diego Health Sciences research on IBD
sciencenewsline.com/news/2016100409450044.html

Social behaviour and the microbiome
sciencedirect.com/science/article/pii/S2352154615001060

UCLA study into how microbiome affects brain function
newsroom.ucla.edu/releases/changing-gut-bacteria-through-245617

Feline microbiome project
kittybiome.com

Dogs and the microbiome
shop.ubiome.com/pages/dogs

Makers of the da Vinci Surgical System
intuitivesurgical.com

Demonstration video of the da Vinci Surgical System
youtube.com/watch?v=_qt5vBW_nu4

First robot cancer operation on lion takes place in Italy
agi.it/archivio/storico/2015/05/08/news/first_robot_cancer_operation_on_lion_takes_place_in_italy-256791

Robot performs soft tissue surgery by itself
popularmechanics.com/science/health/a20718/first-autonomous-soft-tissue-surgery

3D printing makes it to veterinary medicine
avma.org/News/JAVMANews/Pages/140701j.aspx

Dog with 3D printed legs
cnet.com/uk/news/lucky-dog-gets-a-fresh-set-of-3d-printed-prosthetic-legs

Organovo bioprinters
organovo.com

Harvard 3D printing human tissue
seas.harvard.edu/news/2016/10/more-progress-in-building-functional-human-tissues

3D4MD library of downloadable 3D print designs
3d4md.com

The world's first 3D-printed pill is approved
computerworld.com/article/3048823/3d-printing/this-is-the-first-3d-printed-drug-to-win-fda-approval.html

UCL School of Pharmacy 3D printing pills of different shapes
medgadget.com/2015/05/3d-printed-tablets-release-drugs-at-precise-rates.html

The world's first VR surgery
theguardian.com/technology/2016/apr/14/cutting-edge-theatre-worlds-first-virtual-reality-operation-goes-live

IBM Watson diagnosed leukemia that doctors had missed
engadget.com/2016/08/07/ibms-watson-ai-saved-a-woman-from-leukemia

IBM Watson wins Jeopardy
youtube.com/watch?v=WFR3IOm_xhE

Dr. Sofie for vets
veterinarypracticenews.com/LifeLearns-Dr-Sofie-Aims-to-Know-All-the-Answers

AI microscope finds cancer cells more efficiently
newsroom.ucla.edu/releases/microscope-uses-artificial-intelligence-to-find-cancer-cells-more-efficiently

MONITORING AND EARLY WARNING SYSTEMS

Petnostics home urine testing
petnostics.com

Cordio Medical
cordio-med.com

Qualcomm Tricorder XPRIZE
tricorder.xprize.org

University of California San Diego's flexible tricorder
jacobsschool.ucsd.edu/news/news_releases/release.sfe?id=1938

Smart contact lens
computerworld.com/article/3066870/wearables/why-a-smart-contact-lens-is-the-ultimate-wearable.html

Diabetes statistics for dogs and cats
petdiabetesmonth.co.uk/dog_how.asp
petdiabetesmonth.com/cat_how.asp

Smart sutures
economist.com/news/science-and-technology/21702438-turning-surgical-sutures-sensors-all-sewn-up

UC Berkeley sweat sensors
news.berkeley.edu/2016/01/27/wearable-sweat-sensors
Proteus Discover
proteus.com/how-it-works

GOING GLOBAL

I-Vet telemedicine
i-vet.com.au

Cellscope
cellscope.com

The One Health Company
theonehealthcompany.com

CrowdMed
crowdmed.com

VetCompass
rvc.ac.uk/vetcompass

DOES LIFE HAVE TO END?

Michigan State University end of life survey
avmajournals.avma.org/doi/abs/10.2460/javma.242.12.1679

Sooam Biotech
en.sooam.com/dogcn/sub01.html

I spent more than $100,000 to clone my dog
nypost.com/2015/10/31/i-spent-more-than-100000-to-clone-my-dog

Guardian article on dog cloning in South Korea
theguardian.com/science/shortcuts/2013/mar/11/dog-about-to-die-clone-it

First British couple to have their dog cloned
telegraph.co.uk/news/uknews/12133278/First-British-couple-to-clone-dead-pet-dog-pick-up-puppies-from-South-Korea.html

First cat clone turns 10
chron.com/life/article/First-cloned-cat-turns-10-1383844.php

The Cryonics Society of Canada
cryocdn.org

PETS OF THE FUTURE

Boston Dynamics
bostondynamics.com/index.html
telegraph.co.uk/technology/2016/06/24/boston-dynamics-robot-dog-may-be-the-ultimate-household-butler

Human robot relationships
livescience.com/27204-human-robot-relationships-turkle.html

The future of robot pets
hackingtheuniverse.com/singularity/robotics/robot-pets

We may all have robot pets in the near future
huffingtonpost.com/2015/05/14/robot-dogs-pets-replaced_n_7275380.html

Will robots replace real pets?
robotshop.com/blog/en/will-robot-pets-replace-real-pets-in-the-future-16150

Uplifting animals – science or science-fiction?
yalescientific.org/2015/05/science-or-science-fiction-uplifting-animals

Monkeys made smarter using brain implants
io9.gizmodo.com/5943379/for-the-first-time-ever-scientists-have-made-monkeys-smarter-using-brain-implants-could-you-be-next

David Brin on uplifting dogs
davidbrin.blogspot.co.uk/2008/08/unusual-perspectives-uplifting-dogs-and.html

ABOUT THE AUTHOR

Dr Gordon Roberts has had a lifelong love affair with the future of healthcare.

Growing up as a child in New Zealand, he would make regular trips to a nearby forest and, surrounded by some of the Southern Hemisphere's most striking landscapes, he would spend his time wondering what future vet care would look like.

Even then, he had a strong feeling that what we consider to be an accepted standard in healthcare would be replaced by an entirely new paradigm.

He saw a world where the lives of both humans and animals would be greatly improved and extended by the advances to come. This forward-thinking start in life led him to where he is today, one of the world's leading futurist veterinarians with a passion for the medicine of tomorrow.

He is particularly interested in the intersection between conscious energy, medicine and technology. Inspired by exponential advances in human medicine, Gordon is helping bring this knowledge to veterinary science and the broader pet welfare industry.

Gordon divides his time between his native New Zealand (where he lives with his wife, four children and many pets) and conferences, seminars and airport terminals around the world as he spreads the word about futurist vet opportunities, while dealing with his angel investments in future tech.

Want to read more about these exciting developments as they happen? All the latest news and discoveries from the future of veterinary medicine can be found on Gordon's website
futuristvet.com

www.ingramcontent.com/pod-product-compliance
Lightning Source LLC
Chambersburg PA
CBHW040834180526
45159CB00001B/188